Fuels and Petrochemicals Division 2022

Held at the 2022 AIChE Spring Meeting and 18th Global Congress on Process Safety

San Antonio, Texas, USA
10 – 14 April 2022

ISBN: 978-1-7138-6830-9

Printed from e-media with permission by:

Curran Associates, Inc.
57 Morehouse Lane
Red Hook, NY 12571

AIChE
120 Wall Street, FL 23
New York, NY 10005-4020

Phone: (800) 242-4363
Fax: (203) 775-5177

www.aiche.org

Additional copies of this publication are available from:

Curran Associates, Inc.
57 Morehouse Lane
Red Hook, NY 12571 USA
Phone: 845-758-0400
Fax: 845-758-2633
Email: curran@proceedings.com
Web: www.proceedings.com

TABLE OF CONTENTS

641541 Transitioning Toward Net Zero Refining

Abstract:

Today's refineries contribute to Scope 1, 2 and 3 CO_2 emissions because of the fossil crude oil and natural gas they process into gasoline and diesel fuels. Decarbonization of industry and transportation has begun and is expected to continue for the next decades. Neither biofuels (bio-based products) or e-Fuels (based on renewable power) will be cost competitive versus fossil transportation fuels. Both will need subsidies and/or mandates to be commercially viable as did ethanol for road transportation a generation ago. Biofuels and eFuels are being driven by the aviation industry's need for net zero solutions for transoceanic flight, as renewable power and hydrogen are not viable options there. For this reason future refinery configurations will be geared toward jet fuel production, and will be employing new and competing technologies for sustainable aviation fuel.

640802 Tech-Enabled Decarbonization in the Chemicals Industry

Abstract:

Decarbonization is a CEO-level topic for most heavy industrial players. Both external and internal forces are setting the drumbeat to reduce GHG emissions. It is a daunting task that challenging economics behind currently available technologies make only more complicated. The majority of technical solutions that will enable full decarbonization are under development or not even blueprinted yet. Despite these challenges, heavy-industry players need to get going to respond to investor, customers, employees and regulator pressures, while keeping an eye of what breakthrough technologies to invest in for the future. System design, process optimization and energy efficiency constitute the foundations available today to start on the decarbonization path. Artificial Intelligence and Advanced Analytics won't solve all the challenges, but they can certainly provide additional tools required to accelerate the pace and lead the industry pack forward. The chemical industry is home to some good examples of tech-enabled sustainability, yet chemical players can learn from other industries already further ahead in the GHG emission reduction journey. We will provide flagship examples across industries of the "art of the possible" in setting the tech-enabled path to net-zero, and summarize key learning from our work supporting clients in defining and implementing their decarbonization strategies.

639414 Implementation of Feed Forward with PID Controllers

Abstract:

PID controllers are the mainstay in controlling many industrial processes. While the PID controller can provide great performance in most cases, the addition of feed forward control for large measured disturbance can greatly improve the response to the disturbance. This workshop will explain the limitation of PID controller performance, the benefit of feed forward and how to implement feed forward with a PID control scheme. Actual plant examples will be used to illustrate the concepts and results.

640104 Signal Filtering for Control – Why and How

Abstract:

Signal filtering is performed to remove undesirable aspect of a signal, often higher frequency components termed noise. For control purposes, filtering may be needed for effective derivative action and to avoid excessive manipulated variable movement, which can come from proportional and derivative contributions. However, excessive filtering can have a detrimental effect on loop performance leading to sluggish performance or worse, instability. This presentation considers the key reasons to filter a signal, the impact of filtering on key loop metrics, and the types of filters to consider. Illustrative examples are provided to support the recommendations.

639227 Leveraging State Based Control Value in a Program Approach

Abstract:

State Based Control is widely becoming the standard of excellence in the process industry with implementers reporting 80% reduction in operator interventions, along with 90% reduction in alarms. The improved automation has manifested improved safety and environmental performance as well as significantly decreased costly unplanned events. This presentation examines how industry is leveraging successful State Based Control designs in a program approach in multiple geographies over various control platforms

639504 A Simple, Powerful, Binning Approach to MPC Model Gain Conditioning

Abstract:

There is power in the plant testing/statistical regression approach to building dynamic models used in model-based controllers. Using regression to derive a raw APC model will maximize its degrees of freedom, and the result is a model that contains small or non-existent degrees of freedom. Control issues can arise when optimization techniques use these small degrees of freedom to calculate targets for plant operation. Techniques like RGA, SVD, and pivoting are helpful for analyzing control interactions, and conditioning issues can be fixed relatively easily in small models. However, these techniques become difficult and tedious for larger, more complex models. This paper describes a novel, non-iterative binning technique for quickly solving 2x2 conditioning issues for any size model, while guaranteeing gain percentage changes below a certain threshold. Higher order interactions are also discussed.

639384 pH Modeling and Control

Abstract:

Titration curves from labs can be misleading due to sample temperature and composition changes and contamination, poor mixing, split range uncertainties, reagent dilution or substitution, and insufficient data points. Guidance is offered to obtain titration curves that most closely match the process and provide the information needed for control valve size and precision, signal characterization, control system design, and controller tuning.

First principle models can be greatly improved by a universal charge balance solved rapidly by a simple interval halving search for the pH that provides a zero-excess charge. The ability of this universal charge balance to handle a large spectrum of salts, acids, and bases is essential to be able to exactly model the process pH response and corresponding actual titration curve. Theoretical titration curves can have slopes many orders of magnitude larger than the actual titration curves when strong acids and strong bases are used due to omission of conjugate salts and small amounts of carbon dioxide absorption just from exposure to air. High salt concentrations can cause errors due to changes in ionic strength. Means of including these effects in the charge balance are provided. Finally, critical dynamics are included by addition of reagent injection delays, mixing delays, and measurement lags and transportation delays.

pH control can be extremely challenging due to exceptional nonlinearity and sensitivity particularly when strong acids and strong bases are used. Also, electrode life is relatively short even in the best conditions and can decrease to a matter of days due to high temperature and high acid concentration, abrasion, dehydration, and poisoning. Advances in electrode design and signal selection can greatly help. The need to reduce oscillations from valve resolution limit and backlash can be addressed by valve design and use of a PID feature. Finally, the system dynamics and performance can be significantly improved and the cost of the system reduced by better control strategy design and installation.

639347 Robust Batch Process Control Applications: Do's & Don'Ts

Abstract:

While batch processes are somewhat infrequent in the refining and petrochemical industries, they are at the core of many fine/specialty chemicals and, especially, the bioprocess industries. Given the history of the process development work for theses growing industries, the need for fundamental Control & Instrumentation (C&I) engineering has been largely overlooked or, at a minimum, underappreciated. Consequently, when these types of processes are scaled-up from the lab or very small production units to "industrial sized" facilities, they often have difficulties producing at the expected level of performance. One of the root causes for these deficiencies is the improper or poor utilization of control system capabilities and inadequate or improper instrumentation design.

This presentation will provide a conceptual framework for defining and implementing robust and reliable batch controls. Topics addressed include:

- Using the most robust "container"
- Use of built-in control system functionality
- Creating an "automated operator"
- Detecting & Handling Abnormal Conditions
- Sequence Logic Documentation
- Control hardware selection
- Instrumentation Selection

641745 Sponsored Technology Workshop - Reach Your Sustainability Targets with New Levels of Efficiency and Innovation - Aspen Technology

Abstract:

In this workshop, you will learn about AspenTech innovations and how AspenTech University will enable you to maximize business performance and meeting sustainability goals. You will learn how to:

- Reduce use of energy, water and feedstock through process optimization and prioritized capital allocation
- Develop new processes and sources for future energy needs and decarbonize existing applications
- Achieve better Design and Sustainability Outcomes with Concurrent Engineering

639549 Study of Surfactant/Nanoparticles Formulations in Rock/Fluid Interaction for Enhanced Recovery of a Colombian Heavy Crude Oil

Abstract:

Heavy crude oils have been considered as an alternative that could offset the world's energy demand and have been widely studied in recent years. However, this type of crude oil is characterized by being asphaltic, dense, and viscous, with a low API gravity (10 to 22 °API) and high viscosity (>100 cP) in reservoir conditions, as well as its hydrocarbon fractions, tend to adhere to the surface of the rock, causing low mobility compared to conventional crude oil, which generates difficulties in its extraction and transportation processes, making the recovery factor of this type of crude oil low. For this reason, Colombia, occupying fifth place in possession of heavy crude oil reserves in Latin America, has focused on the study of enhanced oil recovery (EOR) techniques to increase the yield of these crude oils in oilfields in a more economically accessible way.

Currently, hybrid techniques such as injection of surfactant/nanoparticle formulations are being studied due to the properties of each of their components. In the case of surfactants, these are used to increase the viscosity of the injection water, making the mobility ratio between water and crude oil increase, but unfortunately, they tend to adsorb to the rock generating a low efficiency in crude oil recovery. To ameliorate this problem, it has been found that dispersions with nanoparticles (NPs) can modify the wettability of the porous medium and transport the surfactant to the oil/water interface instead of being adsorbed to the rock, making the oil encapsulated in the rock more easily released by forming highly stable emulsion with low viscosity and oil/water interfacial tension.

In this study, the effect of formulations composed of a surfactant *S* with silica (SiO_2-NPs) or alumina nanoparticles (Al_2O_3-NPs) on the recovery of a Colombian heavy crude oil of 16 °API @ 25 °C in a sandstone-type rock was evaluated. For this purpose, detergency tests were performed at 60 °C for 120 min, where solutions containing 10,000 ppm NaCl and 5,000 ppm surfactant S were used in the presence of NPs with concentrations in the range of 0 to 1,200 ppm. Finally, contact angle tests were performed to determine the change in rock wettability after the use of the formulations.

The results showed that the recovery percentages of the crude oils in the presence of the surfactant were in the order of 43 - 77 % w/w, while the use of the surfactant formulations with NPs promoted recovery yields in the range of 63 - 97 % w/w. Similarly, it was found that the rocks changed their wettability from oil-wet to water-wet after interacting with the formulations. Finally, the effect of the SARA fractions of the crude oil used on the mineral surface was studied, determining that resins and asphaltenes present a greater interaction with the porous medium, but with the application of this technological innovation rock/fluid interaction is reduced, suggesting that surfactant/NPs formulations can enhance crude oil recovery with a high potential for scaling up.

Based on the above, it has been proven that nanotechnology has entered the oil industry providing multiple benefits such as altering the rock wettability and reducing the oil/water interfacial tension, which allows increasing the recovery percentage of heavy crude oils, thus understanding its importance in EOR processes.

639553 Analysis of Catalytic Aquathermolysis on the Physicochemical Properties and Maltenes-Asphaltenes in Colombian Heavy Crude Oils By Using 1H-NMR

Abstract:

Catalytic aquathermolysis is an enhanced oil recovery (EOR) technique that uses steam injection with a catalyst to produce crude oils in reservoirs with lower energy requirements and less time than conventional methods. Therefore, the crude oils obtained are of better quality due to the new physicochemical properties and composition of maltenes (saturates, aromatics, resins) and asphaltenes. In this work, the catalytic aquathermolysis process was studied in two Colombian heavy crudes. The reactivity tests were carried out in a batch microreactor at 270°C and saturation pressure of 800 psi, for 66 hours, using iron naphthenate at 100 ppm with respect to the metal ion. The use of the catalytic aquathermolysis technique reduced the gas yield in the range of 1 to 3 % w/w, the viscosity from 35 to 60 %. Additionally, distillation curves (ASTM D7169) of the improved crudes showed conversions of the 525°C+ heavy fractions between 4 and 8 %, showing a significant decrease in complex fractions with high boiling points.

Analysis of the average molecular parameters (AMPs) via 1H-NMR spectroscopy confirms that the reactivity and products yield vary according to the complexity of the molecular structure of the maltenes and asphaltenes present in the heavy crude oils. The average molecular parameters showed that the major differences concerning the base crude were aromaticity and alkyl chain length. During the application of this recovery method, it was established that the selectivity and reactivity depends on the composition of the heavy crude oil. Finally, the improvement of the physicochemical properties of crude oil at reservoir conditions is a complementary recovery mechanism to the conventional steam injection process, and in turn, allows the transformation of complex fractions towards the formation of species of smaller size and molecular weight.

639575 Application of Liquid Products from Polystyrene Conversion As Pour Point Depressants in a Colombian Paraffinic Crude Oil

Abstract:

Saturated hydrocarbons known for their difficult treatment due to the content of paraffin composed of long carbon chains from C18 to C60 require a complex study where properties such as crystallization point, viscosity and pour point are analyzed; these are important when studying the nature of the crude oil and must be controlled from the reservoir to the surface. In the absence of an adequate study, these molecules are often deposited when the temperature at which they are exposed is equal to or lower than their crystallization temperature, causing additional costs and large-scale operational problems. In response to this problem, the use of scrapers or heaters has been useful to avoid the agglomeration of paraffin, but these conventional techniques are not completely cost-effective. Therefore, research on recent technologies on the use of pour point depressants (PPD) is gaining importance, as they improve the flow of crude oil easily and effectively by lowering its pour point, preventing the formation of paraffin deposits. The proposed alternative considers the degradation of plastic products by the pyrolysis method and its use as a pour point depressant in paraffinic crude oils, for which polystyrene was disintegrated at 425° C and based on pour point and viscosity tests, its effectiveness was evaluated at different concentrations in two crudes from the Colorado Field, known for its latent problem of precipitation and formation of paraffin deposits. The liquid products obtained from the pyrolysis of the base crude were able to lower the pour point of the base crude by 3 °C with a minimum concentration of 0.5% by volume. The results determined that toluene and the liquid products obtained from the pyrolysis of polystyrene could be postulated as depressants, provided that the corresponding analysis of their application is carried out and the conditions for their use are studied in greater depth

638929 Effect of Initial Water Saturation on Carbonate Rock Wettability: A Contact Angle Study

Abstract:

Carbonate reservoirs hold more than half of the world's remaining oil. Wettability has been proven to have a significant influence on oil recovery from carbonate reservoirs. In most cases, carbonate rock tends to be intermediate-wet or oil-wet due to positive surface charge that induces the adsorption of organic materials. Besides factors including oil composition, temperature, rock mineral composition and so on, the resulting wettability of rock is also affected by initial water saturation. The main objective of this work is to find out how initial water saturation affects carbonate rock wettability.

Indiana limestone core plugs were cut into slices for the purpose of contact angle measurement. Filtered crude oil from the field was used for the oil-aging process. An oven was used to raise oil temperature to accelerate the oil-aging process. A vacuum pump and a desiccator were used to saturate the slices with water or oil. A pressure-plate apparatus was used to establish different initial water saturations. Ambient contact angle was measured using the device Attension Tensiometer.

Two similar curves describing the relation between initial water saturation and the applied air pressure were obtained. Both show a rapid saturation reduction from 100% to 80% in the beginning and a minor reduction as air pressure went above 40 psi. The lowest water saturation value was around 41%, while the air pressure was 66 psi. Different initial water saturation values resulted in different wettability, especially when the oil-aging process was limited to a few days. Generally, samples with lower initial water saturations had a stronger tendency to become oil-wet. After 12 hours of oil-aging, samples whose initial water saturation was lower than 60% became strongly oil-wet. Samples whose initial water saturations were between 60% and 80% became oil-wet. However, samples fully saturated took more than 4 days to become oil-wet. For samples that had high water saturation values, the time required to obtain strong oil-wetness was longer. Samples' oil-wetness was found to be positively correlated to oil-aging time.

This study shows the effect of initial water saturation and oil-aging time on carbonate rock wettability. It also provides a method to have some control on rock slice wettability. This method can be helpful in studies related to wettability.

638931 Influence of Permeability on the Relative Contribution of Wettability Alteration and IFT Reduction in Oil Recovery

Abstract:

Carbonate reservoirs contain a significant fraction of the world's remaining oil. However, carbonate rock tends to be oil-wet as a result of long-time contact with oil in most cases. As a result, wettability alteration is found to be the other major EOR mechanism besides IFT reduction. Some studies have shown that the contribution of wettability alteration is affected by rock permeability. The main objective of this work is to clarify how the rock permeability affects the relative oil recovery contribution of wettability alteration and IFT reduction in carbonate rock. Indiana limestone cores of three different permeability ranges were cut into plugs and slices. They were aged in filtered crude oil under high temperature in the oven for two weeks. A locally synthesized cationic gemini surfactant was used to alter rock wettability towards the water-wet condition. Both ambient contact angle measurement and spontaneous imbibition experiments were conducted for comparison. Results showed that permeability is negatively correlated with the oil recovery contribution of wettability alteration. In spontaneous imbibition tests, lower permeability samples produced a larger fraction of oil in place. This was because capillary pressure helped the imbibition of water phase when rock wettability became water-wet. Lower permeability is related to a smaller pore throat radius and thus implies larger capillary pressure. As a result, in lower permeability rock, the relative oil recovery contribution of wettability alteration was enhanced compared to that of IFT reduction. In the ambient contact angle test, however, slices of various permeability showed comparable contact angles before and after wettability. This observation can be explained by the negligible effect of capillary pressure under such cases. This study shows that when rock wettability is altered from oil-wet to water-wet condition, the contribution of wettability alteration on oil recovery rate is negatively correlated with rock permeability. Besides, this study also suggests the locally synthesized cationic gemini surfactant as an effective wettability modifier. Conclusions obtained in this study indicate that more attention should be given to wettability alteration when optimizing the EOR fluid formula for low permeability carbonate formations.

638932 Development of Magnetic Surfactants for Oilfield Applications

Abstract:

Commercially available surfactants, including polyethylene oxide ether, aryl sulfonate, and petroleum sulfonate, exhibit stability issues when applying high salinity (120,000-220,000 ppm) and high temperature (≥90°C) carbonate reservoir. The severe reservoir environment causes surfactant decomposition and decreases their performance in lowering interfacial tension (IFT), altering rocks' wettability, and mobilization.

For addressing surfactant stability issues, magnetic surfactants were synthesized. The chemical structures were confirmed with the aid of NMR (^{1}H, ^{13}C), FTIR, and mass analysis. The heat stability and aqueous solubility tests were assessed using Thermal Gravimetric Analysis (TGA) and aging methods. The spinning drop method was used to identify interfacial tension. The surface tension and critical micelle concentration (cmc) were measured with a force tensiometer (Biolin Scientific). The rheology was performed on the discovery hybrid rheometer (DHR-3).

The surfactants exhibited good solubility, and the TGA graph showed the decomposition temperature of the prepared surfactants above 230 °C which is superior to oilfield temperature (≥90°C). Similarly, the aging technique revealed no change in the chemical structure after putting them in the oven at 90 °C for one month. The cmc, surface tension at cmc, and interfacial tension of the surfactants were comparable to industrially applied surfactants. According to rheological data, the storage modulus was decreased by enhancing the surfactant concentration at a lower rate of shear and frequency due to charge screening and polymer interaction.

The developed magnetic surfactants exhibited remarkable aqueous and heat stabilities and demonstrated immense potential in severe carbonate rocks.

639300 New Crude Oil Viscosity Models for Burgan Oilfield

Abstract:

In the petroleum industry, the prediction of viscosity using correlations is commonly used for solving many reservoir engineering and surface production problems. In the last seven decades, several black oil viscosity models have been developed. However, these models may only be suitable for the geological regions for which they were developed and may not provide acceptable results for crude oils for other regions.

The purpose of this paper is to create a comprehensive oil viscosity models for the Greater Burgan Oilfield-Kuwait which is the second-largest oil field in the world with estimated recoverable oil of nearly 6% of the world's proven conventional oil reserve. Therefore, 824 data points of oil viscosity measurements for the Greater Burgan Oil Field has been used to evaluate the accuracy of the published models and later used to develop a new oil viscosity models for estimating dead, saturated, bubble point, and under saturated oil viscosities. The validity and accuracy of the proposed models were confirmed by comparing their predictions with 166 measured data points from the field itself.

The new models will improve our predications of oil flow rate, pressure drop in tubing and flow lines and hence the design of production equipment and surface facilities. These models can easily be integrated into reservoir simulation for planning primary, secondary and EOR studies.

640000 CO_2 Dissociation over Ni Supported on Doped Ceria Catalysts

Abstract:

CO_2 dissociation to CO is an interesting fundamental reaction that can be part of the chemical looping process which utilizes natural gas and CO_2/H_2O to generate synthesis gas with solar power. Based on previous research, Ni-$CeZrO_2$ synthesized via various methods, including the combustion methods using citric acid, is a promising material for CO_2/H_2O reforming of CH_4. This study focuses on using temperature programmed reduction followed by a pulse CO_2 dissociation reaction to characterize and evaluate the synthesized catalysts. The reduction characteristics and pulse reaction activity and stability of catalysts are summarized and compared. There are optimum Ni and Zr loadings for the activity, capacity, and stability of Ni-$CeZrO_2$ for CO_2 dissociation. Various pre-treatment methods, such as non-thermal plasma treatments, are studied to modify those catalysts for improved CO_2 dissociation.

640134 Efficient Degradation of Naphthalene Using Hydrazine Assisted Nano-Iron Peracetic Acid Process

Abstract:

To improve the oxidation capability, a well-known reducing agent, hydrazine (HZ) was introduced in the nano-iron/peracetic acid process (NI-PAA). Significant oxidation performance was observed in the HZ-NI-PAA process for the degradation of naphthalene (NAP) in a wider range of pH values (3-9) even with the low dosage of NI i.e., 0.5 μ M. The reactive oxygen species (ROSs) were determined to quantify the presence of hydroxyl radicals (OH) by using electron spin resonance (ESR) technique. Benzoic acid (BA) was used as probe indicator confirming the higher intensity of hydroxyl radicals in the NI-PAA system for the naphthalene degradation. The conversion of Fe (III) to Fe (II) was significantly improved by the facile addition of hydrazine. The decomposition products of hydrazine include nitrogen gas, nitrogen dioxide and nitrous oxide whereas, nitrogen gas was the major environmentally friendly decomposition product among all. Intermediatory products were also determined that proposed three possible different pathways for the degradation of naphthalene. Higher dosages of NI and PAA significantly increased the naphthalene degradation. The humic acid showed inhibitory effect on the degradation of naphthalene whereas, anions (HCO_3^-, Cl^-, SO_4^{2-}) have minor effect. Natural water was used to evaluate the oxidative capacity of Hz-NI-PAA process and further investigations were also performed for the degradation of different micropollutants to find its application. This study significantly improved the NI-PAA process and its possible application in water treatment.

640136 Copper Oxide and Sodium Percarbonate System Supported with Iron Nanoparticles for the Simultaneous Degradation of Trichloroethane and Hexavalent Chromium

Abstract:

Copper oxide and sodium percarbonate system supported by iron nanoparticles was investigated for the simultaneous removal of trichloroethane and hexavalent chromium in the wastewater. The degradation efficiency was evaluated by finding the initial effect concentrations of chromium, trichloroethane and sodium percarbonate as well as the time intervals of sodium percarbonate addition to the aqueous solution. Experimental results confirmed that the removal rates of chromium and trichloroethane reached significantly up to 88.9 and 92.30 % respectively with an optimum dosage of 0.5 mM copper oxide and 1.5 mM Sodium percarbonate. It is presumed that copper oxide and sodium percarbonate promoted the reduction of hexavalent chromium and oxidation of trichloroethane that was completely controlled by iron-nanoparticles. The enhanced degradation of hexavalent chromium is attributed to the availability of the reactive sites of nanoparticles and production of ferrous ions increased the aqueous solution by dissolution and corrosion of nanoparticles. The synergetic role of copper oxide/sodium percarbonate/ iron nanoparticles was observed and identified as this system behaved as an activator for sodium percarbonate as well as reductant for the hexavalent chromium for their degradation purpose. Finally, this system has a great potential for the remediation of chlorinated organic solvents and heavy metals from the wastewater streams.

638294 CO_2 Capture and Conversion to Methanol over a Copper Based Dual Functional Material

Abstract:

The industrial intensification has significantly increased the greenhouse gas emissions, among them carbon dioxide, which is responsible for severe environmental issues, such as global warming and climate change. In this study, a novel bimetallic dual functional material based on 10%Cu/10%CaO/Al_2O_3 was developed to capture and convert CO_2 to methanol. Materials with fixed composition was synthesized in the laboratory and characterized using XRD, XPS, BET, and TGA. CO_2 capture experiments were performed using a magnetic suspension balance up to 20 bars. Meanwhile, conversion tests were performed in a high pressure packed bed reactor at 300 °C and 40 bars. The obtained results in term of adsorption capacity revealed a high CO_2 uptake. While CO_2 conversion tests showed a higher selectivity towards methanol and a good conversion rate.

639224 Research on the Influence of Wax, Resins and Asphaltenes on Wall Sticking of Shale Oil

Abstract:

There is an urgent need to increase the development of unconventional oil resources such as shale oil to ensure the energy security due to the decrease of conventional oil and gas resources in recent years. However, the complicated components such as wax, resins and asphaltenes in shale oil resulting in severe wall sticking and high pipeline back pressure during pipeline transportation. Therefore, it is necessary to study the influence of components on wall sticking of shale oil, which not only provide theory guidance for decreases the back pressure of shale oil pipeline, but also guarantees the safety of pipeline operation.

In this work, the effects of wax, resins and asphaltenes to wall sticking behaviors of shale oil under different temperature, stick-wall time and shear strength are discussed with the help of small scale stick-wall experimental device. To reveal cause of the change in the quality of the wall sticking, the composition and moisture content of sticky wall under different conditions are characterized. Furthermore, based on the theory of nucleation and adhesion, the influence mechanisms of components of shale oil on wall sticking are discussed systematically, so as to advance understanding for the shale oil transportation.

The results show that the lower of temperature and longer of sticking time, the higher of the wall sticking quality. The applying of shearing will reduce the quality of wall sticking quality. Meanwhile, it is revealed that the different contents in wax, resign and asphaltenes causing the difference in wall sticking quality under the same conditions. The synergy of nucleation of wax molecules and adhesion of asphaltene and resins, contributing to the wall sticking of pipeline. These conclusions provide theoretical guidance for ensuring the safety of the shale oil gathering and transportation pipeline in oilfield.

639284 Electrocoalescence of Water-Oil Mixtures Induced By Corona Discharge

Abstract:

Commercially available methods of oil/water separation are mostly energy inefficient, time consuming, and costly. This study considers a contactless method of electrocoalescence via corona discharge. Using a pin-to-plate configuration and pure/pulsed DC sources, non–uniform electric fields form which generate corona discharges. Applying the corona discharge to oil/water mixture results in a smooth separation of the two distinct phases. Considering the different processing parameters, it is possible to tune the effect of effective forces, *i.e.* electrophoresis or migratory coalescence (EP), dipole–dipole interaction (DDI), etc., acting on water droplets to facilitate the separation process. It is observed that the corona discharge significantly enhances the coalescence in concentrated W/O emulsions (made with silicone oil 350 cSt and 2 wt.% water content). After only 120 seconds of processing, it is observed that the two phases were separated into nearly two distinct phases. In the second part of the study, the effect of corona discharge on coalescence rate of water droplets in silicone oil 3000 cSt were investigated. Depending on the initial distance between the water droplets, coalescence could occur due to DDI for close distances or EP for far away droplets. For the case of DDI, it is shown that pulsed DC fields are preferred rather than pure DC. The coalescence times measured were 7, 3, 6, and 18 seconds for pure DC and pulsed DC with frequencies of 1, 10, and 20 Hz. For the case of EP, it was observed that pure DC performs better with coalescence times of 20, 40, 29, and 80 seconds for pure DC and pulsed DC in frequencies of 1, 10, and 20 Hz.

638947 The Blue and Green Energy Translation

Abstract:

The presentation will detail the current trends in green and blue Hydrogen, ammonia and methanol for the new energy transition from fossil to blue and green energy sources. We will also discuss the production technology options for each product. The presentation will also discuss the advantages of Solid Oxide Electrical Cell technology (SOEC) over Alkaline electrolysis for the production of green H2.

SPRING22 + 18TH GCPS

A Joint AIChE and CCPS Meeting

What did I actually buy—Guarantees of Origin For Hydrogen

Achim Wechsung
Exponent, Inc.
1075 Worcester Street, Natick MA 01760
awechsung@exponent.com

Christopher Buehler
Exponent, Inc.
420 Lexington Ave., Suite 1740, New York, NY 10170
buehler@exponent.com

Prepared for Presentation at
American Institute of Chemical Engineers
2022 Spring Meeting and 18th Global Congress on Process Safety
San Antonio, TX
April 10 – April 14, 2022

What did I actually buy—Guarantees of Origin For Hydrogen

Achim Wechsung
Exponent, Inc.
1075 Worcester Street, Natick, MA 01760
awechsung@exponent.com

Christopher Buehler
Exponent, Inc.
420 Lexington Ave., Suite 1740, New York, NY 10170
buehler@exponent.com

Keywords: Green Hydrogen, Guarantee of Origin, Hydrogen Economy

Abstract

For many different products, a buyer cannot distinguish if a claim such as "green hydrogen" or "renewable energy" is true by inspecting the product. For electricity, it was realized that proof of generation by a renewable or green energy source is an essential building block for market success. As a result, Renewable Energy Certificates (RECs) were conceived. A similar mechanism for products such as "blue" or "green" hydrogen enables producers to capture the associated premium. This is especially important in the chemical industry due to its value chains that often span multiple production steps and legal entities. While such certificates were introduced in Europe about 20 years ago for electricity, the EU recently introduced "Guarantees of Origin" for hydrogen. This talk will demonstrate important elements of such certificates and potential pitfalls.

1 Introduction

Over the past decade, consumers, investors, financial institutions, and regulators have increased the pressure to use sustainable and greenhouse gas (GHG) free energy sources in processes and products. This led to renewed interest in the hydrogen economy and, specifically, efforts to commercialize carbon-neutral hydrogen production methods and technologies across sectors.

1.1 The Colors of Hydrogen

Presently, the most widely used production method, steam methane reforming (SMR) followed by the water-gas shift reaction, converts methane from natural gas into hydrogen and carbon dioxide. While the "grey" hydrogen produced in this manner comes with a heavy carbon footprint, this method also lends itself directly to carbon capture and storage. The resulting "blue" hydrogen is anticipated to be the primary method of carbon-neutral hydrogen production in the near-term; however, it is not the long-term goal.

Instead, scientist and engineers continue to work towards producing "green" hydrogen at industrial scale by splitting water in electrolyzers using renewable electricity, e.g., wind, solar, or hydroelectric power. This transition will occur in the coming years, during which time industries and consumers will explore more widespread use of hydrogen. As with the transition to any new technology, we expect the shift from grey, to blue, and then to green hydrogen across industries and applications to introduce new challenges.

1.2 Evolving Hydrogen Economy

Today, hydrogen is mostly consumed on large industrial sites. The main consumers are oil refineries and ammonia plants.[1] Hydrogen is generated on-site or transported to these sites from producers using dedicated hydrogen pipelines such as on the US Gulf Coast or in Belgium, the Netherlands and Germany. Thus, there is a close connection between producers and users.

As hydrogen adoption by users at much smaller scales, such as hydrogen-fueled vehicles or hydrogen-powered residential appliances, increases and hydrogen consumption continues to become more widespread, this close connection between suppliers and consumers may become less evident, e.g., as new intermediaries are added. Tracing hydrogen back to its origins may become near impossible.

1.3 Premium Hydrogen

At the same time, studies show that hydrogen production costs for blue and in particular for green hydrogen still exceed those of grey hydrogen.[2] Thus, consumers of blue and green hydrogen will need to be willing to pay a premium. Many motivations for such a willingness do exist. These can be broadly grouped into three categories: financial, regulatory/compliance and marketing.

Financial reasons include the desire to receive tax credits or a reduction of allocated allowances in an emissions trading scheme. Regulatory reasons can be obligations to purchase, use or sell fuels from lower-carbon sources or obligations to reduce total GHG emissions. Marketing reasons include the ability to sell products into markets or jurisdictions no longer permitting grey hydrogen or the ability to market products with claims such as "green" or "lower-carbon" to consumers.

2 What Color is My Hydrogen

While introducing colors of hydrogen is conceptually useful, even sophisticated hydrogen users cannot distinguish what feedstocks or production technology were used in generating hydrogen. However, for consumers to be willing to pay a premium, they will want some guarantee in return that their purchased hydrogen warrants this premium. Likewise, sellers of blue or green hydrogen want to add credible information to assist them in marketing their product at premium prices.

While international standards on hydrogen quality exists, for example, ISO 14687 or SAE J2719 for hydrogen fuels,[3,4] they are concerned with product specifications such as permitted contaminants and purity. Currently, no accepted international standards exist for blue, green or any other form of lower-carbon hydrogen. Instead, each seller and buyer need to contractually agree on a mutually acceptable definition.

2.1 Contractual Definition for Blue or Green Hydrogen

Arriving at such a definition will require deep technical, regulatory, and commercial understanding of the hydrogen sector. Basic elements that may be included are utilized feedstocks, which could span from coal, natural gas, and other petroleum products to electricity and water, and production technology, which could range from SMR, auto-thermal reformers, and gasifiers to electrolysis or chemical processes that generate hydrogen as a by-product. A contract may also specify an aggregate cap on the GHG emissions associated with the production of hydrogen.

When blue hydrogen is desired, i.e., carbon capture and utilization or sequestration of at least some of the carbon dioxide emissions associated with the production process, parties may decide to specify a lower limit of GHG emissions that must be captured and either utilized or permanently stored.

When the contract specifies green hydrogen, a buyer may request to verify that the hydrogen production is either directly connected to a renewable power production facility or that all consumed electricity has indeed been generated from renewable sources.

When the parties consider other lower-carbon hydrogen, a common understanding of how GHG emissions are calculated is important. Where an electrolysis plant is operated with grid-based electricity, an understanding of the ultimate sources of the grid power may also want to be considered.

After an acceptable definition is established, hydrogen buyers may decide to continuously monitor and verify that the seller meets this definition. Lastly, given the developing regulatory landscape, both parties may want to include a mechanism to respond to changing regulatory requirements.

This process may be feasible for sophisticated sellers, merchants and buyers of large quantities, though the above description suggests that it is unworkable for many other applications.

3 Guarantees of Origin

Overcoming the obstacles outlined in the previous section is possible and other industries provide instructive examples. The power sector faced a similar challenge when renewable electricity entered the grid. Like hydrogen, electricity can be generated using a multitude of "feedstocks" and generating technologies, and customers cannot distinguish these features by inspection of the product. Separately, the chemical industry is confronted with a fundamentally similar issue. Some customers are interested in products made from renewables and are willing to pay a premium. At the same time, the majority of feedstocks will likely remain fossil-based for the immediate future. In addition, chemical processing often involves many subsequent steps with intermediates products. Thus, manufacturers needed to trace the provenance of products through their complex value chains in order to guarantee to their customers that a product was at least partially based on renewables.

3.1 Renewable Energy

The first Renewable Energy Certificates (RECs) were issued in the US in 2001.[5] They are also known as green tags, green energy certificates, or tradeable renewable certificates. In the EU, they are referred to as Guarantees of Origin. RECs represent the technology and environmental attributes of electricity generated from renewable sources.

RECs are usually sold for fixed amounts of energy such as 1 MWh. However, it is possible to separate a certificate from the underlying unit of electrical energy with which the certificate was associated at generation. Then, the certificate can be sold and purchased separately. However, when the certificate is sold separately, the underlying unit of electrical energy is no longer considered renewable to avoid double-counting. This feature provides flexibility to market participants. It is possible to match generators of renewable energy with buyers that particularly value green power but may not be physically connected to such a power source.

3.2 Mass Balance Approach for Renewables-Based Chemicals

Chemical manufacturers use a mass balance approach to conceptually separate renewables-based from fossil-based product quantities. For each processing step, they account for the quantity of renewable resource input to a process and obtain an equivalent amount of product. For example, if the process takes in 1 kg of feedstock A and yields 0.8 kg of product B, then 10 kg of renewables-based material A is converted into 8 kg renewables-based material B. Many production steps are more complicated in the chemical industry, however. Some reasons for this added complexity are (1) more than one feedstock may be needed to produce a product; (2) more than one product is generated in a process; and (3) intermediates are recycled to earlier production stages. As manufacturers perform the mass-balance calculations at each step of their production networks, they are then able to assign which amount of a marketed product was produced from renewables. In the end, an independent auditor certifies both origin and quantity of renewable feedstock and product.[6]

3.3 Guarantees of Origin for Hydrogen

Systems for guarantees of origin for hydrogen work similarly. They allow producers, suppliers and users of green or blue hydrogen to use combined storage and transportation facilities where hydrogen can be combined with hydrogen from other sources.

Figure 1 shows the components of such a system for green hydrogen schematically. A competent body, e.g., a regulatory agency, issues a guarantee of origin for a specific amount of hydrogen when a producer generates this amount hydrogen using renewable feedstocks and energy. The producer can then feed the produced hydrogen into a common hydrogen infrastructure. When a hydrogen supplier delivers the same amount of hydrogen to an end user, the competent body cancels a guarantee of origin for the same amount. In other words, a separate bookkeeping system keeps track of the amounts of green hydrogen and ensures that it cannot be claimed more than once. The physical hydrogen is detached from this guarantee of origin. Thus, it is possible that the end user of green hydrogen is not even physically connected to the same hydrogen system.

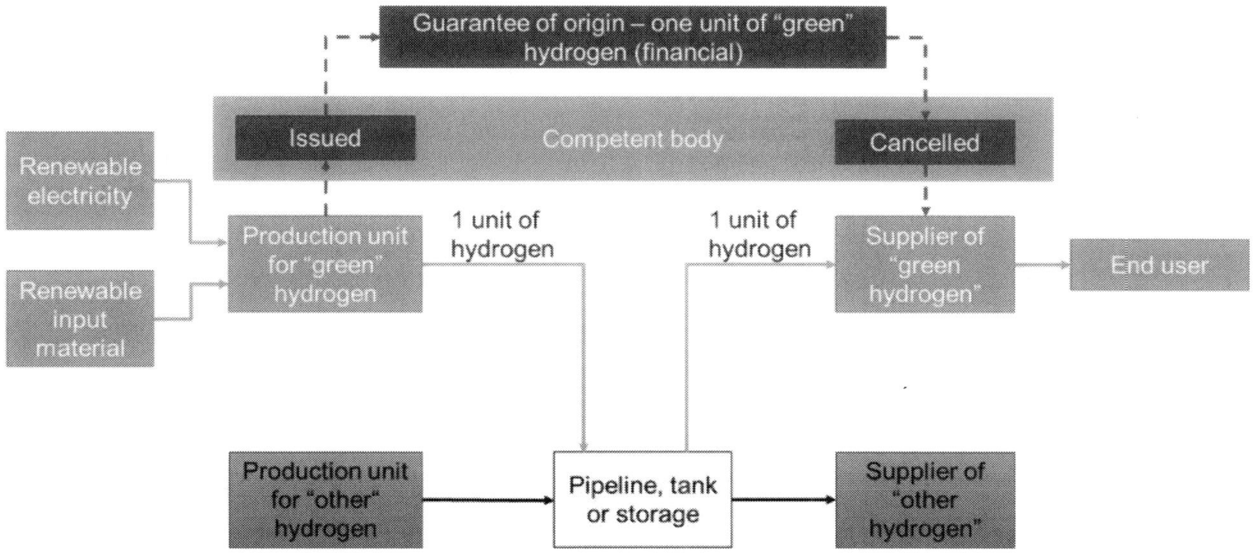

Figure 1. Schematic showing a guarantee of origin system for green hydrogen.

<u>3.3.1 Advantages</u>

Compared to the contractual definitions and complicated verification mechanisms, such a system provides clear benefits. It also relies on experience gathered for similar schemes that certify electricity generated from renewable sources. Furthermore, it resolves the issue of specific, contractual requirements for blue or green hydrogen. And the buyer can rely on the competent body rather than having to verify claims about the seller's organization and supply chain.

It also offers financial advantages. Because the attribute of green hydrogen or blue hydrogen is decoupled from a physical flow of hydrogen, interested parties can trade the guarantees of origin separately from the physical hydrogen. For example, a refinery operator may want to use a certain fraction of green hydrogen on site so that some products can be marketed as renewable. However, the refinery may not be connected to a supplier of green hydrogen. The operator could instead procure a sufficient number of guarantees of origin on the market.

It is important to point out that guarantees of origin are mainly legal or regulatory schemes. Thus, buyers and sellers depend on what has been adopted in their jurisdictions.

<u>3.3.2 Examples</u>

One example of such a system is CertifyH2, a demonstration project for a guarantee of origin system in Europe.[7] It provides definitions for "green hydrogen" and "low-carbon hydrogen". The latter originates from a non-renewable source and may include carbon capture and utilization or storage.

While still at an earlier stage, the Australian government is also working towards establishing a guarantee of origin system for hydrogen, which may include other low-carbon products.[8] In December 2021, it announced that industry trials would be conducted during the next 18 months.

4 Conclusion

Green, blue and grey hydrogen cannot be easily distinguished. However, for a market for green and blue hydrogen to develop, market participants need a common definition. The power sector faced a similar challenge and introduced RECs. Similarly, guarantees of origin for hydrogen can achieve the same. They provide confidence to buyers of green and blue hydrogen and allow sellers to distinguish their product in the marketplace. Lastly, they can also be decoupled from the physical hydrogen delivery so that blue or green hydrogen becomes available at sites that are not physically connected to a producer of blue or green hydrogen.

5 References

[1] International Energy Agency. *The Future of Hydrogen*. 2019. https://www.iea.org/reports/the-future-of-hydrogen. Accessed on March 3, 2022.

[2] International Energy Agency. Hydrogen production costs by production source, 2018. https://www.iea.org/data-and-statistics/charts/hydrogen-production-costs-by-production-source-2018. Accessed on March 3, 2022.

[3] International Organization for Standardization. *Hydrogen Fuel Quality — Product Specification*. ISO Standard 14687. 2019.

[4] SAE International. *Hydrogen Fuel Quality for Fuel Cell Vehicles*. SAE Standard J2719. 2020.

[5] Holt E, Bird L. *Emerging Markets for Renewable Energy Certificates: Opportunities and Challenges*, NREL/TP-620-37388. 2005. https://www.nrel.gov/docs/fy05osti/37388.pdf. Accessed on March 14, 2022.

[6] Scott A. The mass-balance way of making chemicals. Chem. Eng. News 2021; **99**: 15.

[7] Barth F, Winkel T, Vanjoudt W, Kuronen A, Lehtovaata, M, Altmann M, Schmidt P, Veum K, Weeda M, Castro J, Nürnberger K. *Towards a Dual Hydrogen Certification System for Guarantees of Origin and for the Certification of Renewable Hydrogen in Transport and for Heating & Cooling, Final Report of Phase 2*. Fuel Cells and Hydrogen Joint Undertaking 2. 2019. https://op.europa.eu/en/publication-detail/-/publication/3d66be8f-6284-11ea-b735-01aa75ed71a1/language-en. Accessed on March 14, 2022.

[8] Department of Industry, Science, Energy and Resources. *A Hydrogen Guarantee of Origin scheme for Australia*. Commonwealth of Australia. 2021. https://storage.googleapis.com/converlens-au-industry/industry/p/prj1a3de348a6c0ad7d282f7/public_assets/Discussion%20paper%20-%20A%20Hydrogen%20Guarantee%20of%20Origin%20Scheme%20for%20Australia.pdf. Accessed on March 14, 2022.

639263 Selective Hydrogenation of Light Hydrocarbons Products of Flash Hydropyrolysis of Grasses Pastures Using a Fluidized Bed and Pd Ce Ni Eggshell Catalyst.

Abstract:

Selective hydrogenation of light hydrocarbons, produced in a flash hydropyrolysis of grasses pastures, using a fluidized bed reactor and Pd Ce Ni eggshell catalyst.

Roberto Galiasso Tailleur[a,b,c]

[a]Simon Bolivar University, Baruta, Caracas Venezuela and [b]Hypro-Consultant, 4250 Corrine Dr Suite 204, Orlando Florida 32814

[c]roberto.galiasso@hyproconsultant .com

Extended Abstract

Hydrogenation of the fraction C_1-355°C, produced during the flash hydropyrolysis of grasses pastures was performed in a fluidized bed reactor using a thio-resistant PdNiCe on silica eggshell catalyst. The catalyst, in bids of 80 micron of diameter, contains 0.1% of Pd, 0.2% of Ce and 0.3% of Ni on surface. Metals were incorporated on surface by impregnating the gel precursor of the support and then calcining it at 480°C. Before used the catalyst was reduced in hydrogen at the same temperature. The effects of temperature and pressure in apparent rates of olefins and diolefins hydrogenation and catalyst deactivation were measured in 25 short tests and then in 7 day pilot plant continuous operation. The feed was obtained from a high-pressure high-temperature separator (HPHTS) of the inverted hydrocyclone reactor [1] pilot plant that process perennials forage. The gas phase, obtained by the top of HPHTS at 12MPa and 593 K, contains H_2, H_2S, water, NH_3 and C_1-355°C hydrocarbons; it was cooled to either 633, 643 or 653K and depressurized to either 11, 9 or 7MPa before being introduced in a fluidized bed reactor (HDT) to be hydrogenated. Catalyst was added and withdrawn from the bed continuously to maintain constant the activity. The heavy liquid fraction of HPHTS can be used, among other possibility, as feed in the Delayed Coking (DC) unit. The composition of the light-hydrocarbons and those of the hydrogenated products are shown as an example in **Table 1**. The feed to the HDT reactor contains olefins and diolefins with the following distributions: 40 and 36% in the C_4-C_5, 43 and 48% in C_6-C_{10} and 17 and 12 % in the C_{10}-355 °C fraction, respectively. The HDT product obtained at 9 KPa, 543 K, 0.2 and 0.4h of gas and solid residence time shown that there is more than 90 and 56% of diolefins and olefins conversion. Most of none-aromatics sulfur (mercaptans, and other thiols), oxygen (carboxilic acid, ester, and alcohols) nitrogen content hydrocarbons were reduced. Thus, product acid content was reduced as well as the amount of solid formed during the accelerate storage test (one week at 55°C under nitrogen) reported in **Table 1**. The spent catalyst contains an average of 1.1 wt % of coke. Heteroatoms containing hydrocarbons in aromatic structure are not modified by the HDT at this condition. It is used to remove the most reactive olefins and diolefins that accelerate the HDS catalyst deactivation.

Table 3 shows the results of a continuous operation for 8 days at different operating conditions. Each condition was maintain for one day. Notice the effect of temperature, gas residence time and hydrogen partial pressure in olefins and diolefins conversion, as well as the catalyst throughput used to maintain the activity. The mass transfer calculation shows that hydrogenation of diolefins and olefins are not affected by diffusion from gas to external surface of the eggshell catalyst. The reaction produces additional gases, hydrogenated naphtha-diesel fraction (C_5-355°C) and coke. The hydrogenated product needs to be treated in a conventional HDS (hydrotreater) to adjust further the properties and fulfill the fuel specification of the fractions.

Coke is accumulated on catalyst in multiple layers on active sites and on silica support; coke is withdrawn with the spent catalyst and then it was regenerated by combustion; particles have a carbon content distribution equivalent to a perfect mixed flow function (E(θ):1.1 CSTR). Higher the temperature and lower the pressure and the solid throughput, higher the amount of coke accumulated on surface. Nevertheless, not all the coke is deposited on active sites. The probability that coke grows on top of acid sites (Po) increase with temperature between 0.16 to 0.20 (**Table 2**) but is not affected by pressure. Coke dispersion was measured by high resolution electron microscopy and XPS analyses of the surface. The use of silica bids instead of catalyst at the above mentioned three operating temperature shows a higher coke deposition of coke (1.6, 2.3 and 2.6 wt. %) than in presence of catalyst, but there are the same distribution of particle size on surface (between 12-40 microns diameter). Average coke content at the steady state operation with PdCeNi catalyst is 0.82, 1.1 and 1.4 wt. % on the solid withdrawn respectively at 533, 543 and 553 K (9MPa, 0.2 and 0.4h of gas and solid residence times). The solid flowrate was changed according to temperature, gas residence time and pressure of operation used in the plant. The solid throughput employed at different operating conditions was calculated using a model of reaction and deactivation obtained previously [2]. The model consider that deactivation of hydrogenation active sites is dependent on olefins diolefins and hydrogen partial pressure (equation (**1**)); the apparent kinetic rates are based on olefins and diolefins disappearances with deactivation. Olefins and Diolefins are hydrogenated on the solid (adsorbed species), but they are polymerized in gas phase producing nanoparticles of coke that deposit on catalyst; coke rate of built-up depend upon olefins and diolefin olefin and hydrogen in gas phase (see equation in (1); catalyst activity (*a*) depend upon amount of coke formed and the probability of deposition on active sites at the external surface (0.09 m^2/g); Rate of reaction and deactivation were solved using a Runge-Kutta and Genetic Algorithm to obtain the values of the rates constants; they are needed to predict catalyst throughput at different operating conditions.

The results obtained during 7 days of continuous operation, displayed in **Table 3,** confirmed the effects of variables in activity and the stability of the operation. There, the effects of variables in olefins diolefins, sulfur and hydrogen consumption can be seen in the rows. Notice that the system achieved the same level of conversion at the beginning and at the end of tests thank to the use of adequate throughput of catalyst mentioned in the table. The information obtained was employed to confirm the values of the activation energies that are reported in **Table2.** The table shows that activation energy for catalyst deactivation (E_d) is 21 and 35% higher than those of hydrogenation of diolefins (E_{dol}) and olefins (E_{ol}). Deactivation occurred by coke coverage of active sites. The XPS analyses of coke and Ni content on surface indicated the metal dispersion decrease with the logarithm of coke content. HDT stage is operate at low temperature to minimize the gas phase polymerization of the most reactive olefins and diolefins. The hydrogen consumption is low at this conditions and the catalyst can be regenerated.

[1]Galiasso Tailleur R. Gonzalez Y, Lucena M.;"New invertd cyclone reactor for flash pyrolysis

(2014) Catalysis Todays 220-222,186-197.

[2]Galiasso Tailleur R. "Hydrogenation and desulfurization in gas phase of light hydrocabons from hydrocracking and delayed coking. (2019) Catalyst deactivation. Chem. Eng. Sci. 210,3 115-195

r_{Ol}(mol/gh)=$k_{Ol}.P_{H2}^{0.3}.P_{Ol}^{0.2}.a_f$; r_{Diol} (mol/gh)=$k_{Diol}.P_{H2}^{0.3}.P_{Diol}^{0.3}.a_f$;r_{coke} (wt%/h)=$k_{coke}P_{Ol}^{0.3}.P_{Diol}^{0.28}.(0.28+PH2)^{-0.12}$; r_d (a/day)=$k_d.w_{coke}.Pr.a$; t_s(Lh/L):$(1-a_f)/(k_d.a_f)$; $a_f=1/(1+r_d)$ (1)

Table 1 Feed and Prod. Prop.			**Table 2 Activ.**		**Table 3 Pilot plant operation**								
Feed wt. %	**Feed**	**Prod**	**Energy Kcal/mol**		**Days**	**1**	**2**	**3**	**4**	**5**	**6**	**7**	**8**
Sulfur	3.4	2.9	E_{ol}	22	Diol %	96	83	77	79	76	79	75	95
Oxygen	0.06	0.042	E_{dol}	19	Olef %	82	76	68	78	73	70	68	81
Nitrogen	0.04	0.035	E_{coke}	22	Cat.g/h	0.6	0.55	0.5	0.56	0.54	0.56	0.51	0.6
Olefins PiONA	23.2	7.2	E_d	28	T K	553	543	533	543	543	543	543	543
Diolefins Mav	26.0	<1	*Po* (prob.)	0.18	P MPa	9	9	9	11	7	9	9	9
Acidmg KOH/g	22.0	5	RTD E(θ)	N:1.1	t_g (h)	0.4	0.4	0.4	0.4	0.4	0.3	0.2	0.4
Ac. Stor. mg/g	4.2	0.14	Coke HRTM	20-48	S %	2.7	2.9	3.0	2.8	3.1	3.0	3.2	2.7
C2 – 355 ^oC	96.3	95.8	H_2 consumption mol/g			1.2	0.9	0.7	1.1	0.8	0.7	0.8	0.9
C1	0.8	0.9											

638807 Radio Frequency Driven Heating of Catalytic Reactors for Portable Green Chemistry

Abstract:

Here, we propose a novel integration of radio frequency (1-300 MHz) responsive nanomaterials with conventional catalytic materials to realize a new class of heterogeneous catalysts that undergo uniform volumetric and localized heating to drive chemical transformations at the modular scale. Approximately 80% of chemical manufacturing involves heterogeneous catalytic reactions, which currently require heating via steam utilities or fired furnaces, and thus contribute to global greenhouse gas emissions while also limiting distributed chemicals production. Our approach uses an electric route to produce chemicals where radio frequency (RF) electric fields and their interaction with carbon/ceramic nanomaterials are utilized to selectively heat the catalyst composition. A proof-of-concept is demonstrated using the commonly studied methanol steam reforming reaction on a platinum catalyst. In this study, two RF susceptors were used: carbon nanotubes and silicon carbide fibers. The conversion rate of methanol using RF heating was comparable to oven heating at varying temperature and catalyst combinations. This is a potential breakthrough over conventional catalytic reactors in that it enables small, safe, sustainable, on-site, and on-demand production of chemicals in the absence of traditional manufacturing infrastructure.

639330 Hydrotreating of Bio-Oil Model Compounds: Experimental Results and Kinetic Modeling

Abstract:

Introduction

Although bio-oil from fast pyrolysis of lignocellulosic biomass is a potential source for fuels and valuable chemical building blocks, such bio-oils cannot be directly integrated into existing petroleum refining facilities. Indeed, fast pyrolysis bio-oil has an extremely complex organic character with not only a high amount of carbon, but also large quantities of oxygenated species. Since the presence of oxygen reduces the quality of bio-oils as a fuel source, bio-oils require a dedicated pre-refining step, which is generally a hydrotreating process. However, due to its high complexity, in both molecular and reaction terms, the upgrading of bio-oil reveals to be difficult. To correctly design such a process, one must acquire insight in the molecular composition of bio-oils and of the reactivity of the various species.

The present work describes a methodology for modeling the upgrading of bio-oils. The proposed method will be applied to construct kinetic models for the hydrotreating of bio-oil model compounds. This strategy will enable a further understanding of bio-oil reactivity, opening the doors for the modeling techniques for full-range bio-oils.

Methodology

As modeling strategy, a direct discrete simulation of reaction events was chosen. The approach is based on a kinetic Monte Carlo (kMC) method, termed Stochastic Simulation Algorithm (SSA)[1,2], and describes the evolution of a system, molecule by molecule. Unlike the classic procedures, this method can describe the evolution of a molecular mixture over time without a pre-defined reaction network.

The stochastic approach is based on the identification of all possible reaction events at a given moment in time and on the calculation of the reactivities for each event. Thanks to the work of Gillespie[1,2] and recalculating apparent first order rate coefficients for the reactions, the reaction probability can be generated for each reaction event. By assembling all the probabilities of all molecules, the overall reactivity allows to determine, via Monte Carlo sampling, the time interval until the next reaction, and to select a given reaction. The sampling process of both variables is then repeated until the final simulation time is reached.

The SSA requires as inputs a list of reactant molecules, the different types of reactions that can occur and the kinetic parameters associated to these transformations. Due to its stochastic nature, this method generates a different outcome for each simulation, which does not allow to construct the general evolution of the reaction system over time. Therefore, as with any stochastic approach, the method has to be based on several simulations, each of which calculates discrete time trajectories of molecular populations. Their average profile will converge to the same results as deterministic methods.

Results

To understand the reactivity, the hydrotreating of two model molecules was studied: Guaiacol and Furfural. The experimental data was obtained by Ozagac[3] and by Costa da Cruz[4] for the hydrotreating of guaiacol and furfural over a $NiMo/Al_2O_3$ catalyst in a batch reactor at 13 MPa, at 200°C, 250°C, and 300°C, and for reaction times of 3h and of 5h. For several experimental conditions, blank runs were performed and showed a relatively high conversion. For guaiacol, an initial conversion of 10% was obtained at 300°C after the heating up and cooling down the batch reactor.

For the kinetic modeling of hydrotreating guaiacol and furfural, the reactions for hydrogenation, dehydrogenation, demethylation, demethoxylation, decarbonylation, decarboxylation, deoxygenation, transalkylation, and dehydration were implemented in the Stochastic Simulation Algoithm. For the model molecules, the SSA method is able to predict the expected trends, not only for the conversion of the reactants, but also for the generation of the various products and for the hydrogen consumption. Additionally, this approach also predicts the transformation of the several reaction products into smaller molecules. Due to the high reactivity of guaiacol, the heating and cooling periods contribute significantly to the overall conversion. Hence, the reactor simulations need to account for these dynamic periods.

The model results were compared to the experimental data. Accounting for the thermal behavior of the reactor during heating, reaction and cooling, the kinetic parameters of the reactions were estimated. A good agreement is observed. The comparison of the simulated and the experimental selectivities for methanol and dialcohols shows that the model predicts, within the error of the experimental data, the correct order of magnitude for both chemical families.

Conclusions

The SSA methodology was applied to bio-oil model molecules and showed the correct trends when compared to available experimental data. Through this study, a kinetic model was generated that is able to simulate the reactions of guaiacol and furfural under hydrotreating conditions. It was shown that the SSA method is an excellent simulation tool for complex reaction networks without the need for a pre-defined reaction network. Furthermore, this method enables a better understanding of these systems, making it suitable to be applied to a full range bio-oil.

Literature Cited

1. Gillespie D.T., "A general method for numerically simulating the stochastic time evolution of coupled chemical reactions," *Journal of Computational Physics*, **22**(4), pp. 403–434 (1976).
2. Gillespie D.T., "A rigorous derivation of the chemical master equation". *Physica A: Statistical Mechanics and its Applications* **188**(1-3), pp.404–425 (1992).
3. Ozagac M., "Etude mécanistique de l'hydroconversion catalytique de bio-huiles de pyrolyse," PhD thesis, Université Claude Bernard - Lyon 1 (2016).
4. Costa da Cruz A.R., "Compositional and kinetic modeling of bio-oil from fast pyrolysis from lignocellulosic biomass", PhD thesis, Université Claude Bernard - Lyon 1 (2019).

639344 Computational Characterization of Pore Network Models for Gamma-Alumina Catalysts

Abstract:

Introduction

Understanding how transport phenomena and chemical reactions within porous catalysts are affected by the texture of the support opens the door to optimizing the catalyst performance. New and more rigorous heterogeneous models for chemical reactor design could be created by accounting explicitly for the phenomena occurring at the pore scale. To model structures of porous catalysts several approaches can be found in literature [1,2]. Generating computational structures that match structural descriptors of the real material requires information about the textural properties and the topology of the material. In this work, the impact of the porous structure of gamma-alumina catalyst supports on mass transfer phenomena in non-reactive and reactive systems is studied. The representation of the alumina support start from previous work [3], in which a network of interconnected cylindrical pores is created by means of a stochastic assembly algorithm. By using a hierarchical network, Ferreira [4] was able to computationally generate several pore network structures whose textural and transport properties matched those obtained experimentally in Kolitcheff's work [5]. The objective is now to further constrain the number of possible structures by generating a network that allows representing experimental results obtained from different characterization techniques. Nitrogen Porosimetry (NP), Mercury Porosimetry (MP), Nuclear Magnetic Resonance Cryo-Porometry (CP) and Thermo-Porometry (TP) are such characterization techniques that deliver quantitative and qualitative information about the porous structure by using theoretical and empirical models: pore size distribution, pore volume, and surface area are textural properties estimated by these techniques [6].

Methods

The pore networks used for the simulations were generated through a discrete reconstruction model based on a stochastic algorithm described by Ferreira [3,4]. For the models, two cases are considered: in the first case, the pores are modelled as cylinders that are connected through zero volume nodes, while in the second case the pores are cylindrical and connected to spherical nodes. The cylinders are organized in space using a 3D cubic lattice that can be regular or disordered. By using a Monte-Carlo approach, the creation process of every pore is totally random, constrained just by the pore existence probability defined as input.

To characterize the digital pore network, simulation algorithms for NP, MP, CP and TP were developed. The algorithms account for percolation phenomena and confinement effects occurring within the porous network during the characterization. Selection of the right computational algorithms to model the physical phenomena and to store and handle the network's information is quite relevant to limit the execution time. An Invasion Percolation (IP) algorithm was adapted for each modelled technique based on the work of Wilkinson and Willemsen [7] and the improvement made by Masson and Pride [8].

Finally, the pore network model representation of a gamma alumina has been used for simulating intraparticle diffusion in the case of selective hydrogenation of butadiene.

Results and discussion

The textural properties of the digital pore network were calculated both geometrically and by processing the results of the simulated characterization techniques. Compared to the direct calculation of the textural properties by simply counting the numbers of pores, the direct simulation of the entire characterization profiles provides a better interpretation of the experimental results by analyzing the validity of the assumptions made by theoretical and/or empirical models used for the calculation of the material's textural properties from the experimental data. The pore network generation tool is based on a ***O(log(n))*** algorithm that allows to create a network with pores in only 46 seconds on a Dell desktop computer with a 3.5 GHz Intel Xeon E5 CPU and 16GB of RAM. For the characterization techniques, the use of a dynamic priority list to handle the exploration of the network is valid for all techniques, as the phase change and/or percolation phenomena occurring in every technique are constrained by the topology of the network. The algorithms are capable of reproducing typical hysteresis behavior observed in the experimental results for alumina catalyst supports. By versioning the IP algorithm to our needs and handling the priority list using diverse versions of binary heaps, we were able to maintain a good time performance for the different algorithms created for each characterization. The execution time for the characterization techniques corresponds to an ***O(nlog(n))*** behavior and takes about 25 seconds for pores. To match the textural properties of the digital pore network with that of an actual alumina support, the input variables of the network model are modified until the differences between the output textural properties and actual values are minimized. A good agreement was observed between the properties of the digital twins and the actual aluminas.

Using the pore network model representation, the intraparticle diffusion was simulated in the case of selective hydrogenation of butadiene. By varying the pore network properties, their effect on the effectiveness factor was investigated.

Conclusions

In this work, a pore network generation algorithm was used to represent gamma-alumina supports. To obtain the textural properties of such a digital pore network, an Invasion Percolation algorithm was adapted in order to characterize this structure by means of the computer-equivalent of nitrogen porosimetry, mercury porosimetry, NMR cryo-porometry, and thermo-porometry. Using the developed algorithms, it was possible to generate digital pore networks whose properties are in good agreement with the properties experimentally measured by different characterization techniques on actual gamma-alumina supports. Finally, the pore network model representation of a gamma alumina has been used for simulating intraparticle diffusion in the case of selective hydrogenation of butadiene.

References

1. Ye G., Sun Y., Zhou X., Zhu K., Zhou J., Coppens M.O., Chemical Engineering Journal, 329 (2017) 56-65.
2. Euzen P., Raybaud P., Krokidis X., Toulhoat H., Le Loarer J.-L., Jolivet J.-P., Froidefond C., In Handbook of Porous Solids, Vol. 3. Weinheim: Wiley-VCH Verlag GmbH, 2002, 1591-1677.
3. Carreira Ferreira S., 2018, Modelling of the intra-granular mass transfer within catalytic reactors. Université de Lyon (in English).
4. Ferreira S., Verstraete J., Jolimaitre E., Leinekugel-le-Cocq D., Jallut C., Computer-Aided Chemical Engineering, 40, Part A (2017) 91-96
5. Kolitcheff S., 2017, Approche multitechnique des phénomènes de diffusion en hydrotraitement de distillats. Université de Lyon.
6. Coasne B., New Journal of Chemistry, 40 (2016) 4078-4094.
7. Wilkinson D., Willemsen J., J. Phys. A: Math, 16 (1983) 3365-3376.
8. Masson Y., Pride S.R., Transport in Porous Media, 102 (2014) 301-312.

638444 The Active Site and Mechanism for Catalytic Hydrogenation on Metal Cation Catalysts Supported on Nu-1000: Insights from Experiments and Microkinetic Modeling

Abstract:

The mechanism of ethene hydrogenation to ethane on six dicationic 3d transition metal catalysts is investigated. Specifically, a combination of density functional theory (DFT), microkinetic modeling, and high throughput reactor experiments is used to interrogate the active sites and mechanisms for Mn@NU-1000, Fe@NU-1000, Co@NU-1000, Ni@NU-1000, Cu@NU-1000, and Zn@NU-1000 catalysts, where NU-1000 is a metal–organic framework (MOF) capable of supporting metal cation catalysts. The combination of experiments and simulations suggests that the reaction mechanism is influenced by the electron configuration and spin state of the metal cations as well as the amount of hydrogen that is adsorbed. Specifically, Ni@NU-1000, Cu@NU-1000, and Zn@NU-1000, which have more electrons in their d shells and operate in lower spin states, utilize a metal hydride active site and follow a mechanism where the metal cation binds with one or more species at all steps, whereas Mn@NU-1000, Fe@NU-1000, and Co@NU-1000, which have fewer electrons in their d shells and operate in higher spin states, utilize a bare metal cation active site and follow a mechanism where the number of species that bind to the metal cation is minimized. Instead of binding with the metal cation, catalytic species bind with oxo ligands from the NU-1000 support, as this enables more facile H_2 adsorption. The results reveal opportunities for tuning activity and selectivity for hydrogenation on metal cation catalysts by tuning the properties that influence hydrogen content and spin, including the metal cations themselves, the ligands, the binding environments and supports, and/or the gas phase partial pressures.

AUTHOR INDEX